POVERTY IS A SIN
NOT A
GUARANTEE TO SIN

Albert Yaw Ofori

ALBERTOLOGY

For more information or bulk purchase, contact:

Phone number: +233 (0) 241494233
Email: albertofori16@yahoo.com
Facebook: Albertology
Twitter: @albertologic

ISBN: 978-9988-2-8245-5

CONTENTS

PREFACE

This book is aimed at providing a detailed knowledge to individuals preparing for the future. It is also useful to persons of all caliber. It can help you-

- Build your mental capacity.
- Stand adversaries.
- Live above ordinary.
- Make right decisions.
- Think positively.

In one way or the other, everyone is a thinker depending on what is being thought. Being a thinker is a great opportunity and responsibility. The most important thing you will ever think are the principles of life as far as the earth is concerned. Poverty has become a major challenge in our generation and that needs to be rectified. This book is something else, maybe it is not just a book but a mystery. It comes with some quotes which have the power to deliberate the minds of both the poor and the rich. Due to the mysterious nature of the book, you need revelation and continuous reading to fully understand the whole concept.

This masterpiece has been written based on raw facts, experiences and revelations. It is a child centered, young centered and adult centered book which has been prepared on the scope of mysteries. The best way to vividly understand this masterpiece is to read with both your eyes and your mind. Eye reading will only interest the eye and mind reading will surely pave way for change. If you can read this book without distraction, you are a genius. The undiluted perception about this book is to activate your thinking capabilities and as you think, you shall take and as you take, you shall change.

The primary aim is to change people especially the young and the generation yet unborn. If you want the world to be under your feet, this is your book. If you want to swim in the ocean of power, this is your book. If you want to prosper in all areas of your life, this is your book. If you want to live a happy life, this is your book. If you want to be extraordinary, this is your book and if you truly want to be that change agent, then your book is this indeed.

The value of all humans from God's perspective is the same but the value of all humans from man's perspective is always not the same because of poverty. We now have people with class, people as kings, the rich, the poor and people as slaves all because of the way we do things on earth. Since the rise of poverty, things have never been the same due to some taste of preferences. Some people might place high value on clothes, shoes, cars, mansions and money whiles others might place such value on other things. All these (material) things are good but its goodness should never and ever override the goodness of mankind.

Your life will never be the same after reading this book because every word or statement has its own impartation. The more you take a second look, the more the impartation. To do this effectively, you must log on your brains as we discover mysteries pertaining to future prospects.

THE ROOT OF POVERTY

CHAPTER 1

There are a lot of definitions concerning poverty. It is generally defined as the state or condition of having little or no money, goods, or means of support, general scarcity or the state of lacking certain amount of material possessions. I can categorically state that, poverty is the state of desperation, inability to possess possessions and the total confusion in the mind of those who lack the ability to change situations.

The Earth was borne out of prosperity. Everything was sinless and without fault. The ideas, the word, the vision, the trees, the sea, the land, were all perfect. Man was introduced and he became the center of all these flawless nature. He was given the power to define, initiate, co-ordinate and to create new things with the mind. The mind became the center of man and man became the center of the Earth. Man is mind.

Everything was rich and man was made rich. Poverty was not in the dictionary of mankind and it was totally an outcast. Mankind discovered, created and developed new things with the mind. The mind began to differentiate between good and evil, right and wrong, sweet and sour, blind and sight, poverty and prosperity. Poverty was decoded after man realized a certain kind of life below ordinary. Prosperity was reigning and mankind decided to use the mind to degrade all the beautiful system nature offered.

Man became totally perplexed and decided to test certain things which was below ordinary. The beautiful system nature offered was not appreciated and therefore, mankind decided to taste different kind of life. Upon series of discovery, the human mind was divided. Confusion set in and man became vulnerable to so many things and poverty had it way. A confused mind is a poor mind and a poor mind is a confused world.

When the mind suffers, the world suffers. Poverty had it way from the mind and became the subject of the day. The world was initiated into poverty through the mind and gained grounds. The human race took different dimension. People developed different concepts, different ideas, different visions, different taste, different attitudes and different mindset.

Prosperity became the language of class and poverty was subordinated with canker. The battle line was drawn and prosperity and poverty became enemies. The rich saw the poor as odd and vice versa. The language of a rich man is never understood by a poor man.

The world was disintegrated and certain norms were spelled out. The poor became an alien in a rich world, the smart world was distinguished, the brave was acknowledged, fools were despised and thinkers were established. In the natural sense, every human has his or her own world. Making it to a different world is a decision and sometimes by choice. Every human has the power to enter any territory as far as the world is concerned. Every world has a security gate and the only key to unlock such a world is the mind.

There is no way you can build a house without the thought of building. You will definitely have some level of desire and desire is of the heart and the heart is of the mind. The mind controls your life and whatever life you live is the result of your mind.

Poverty is just like a lion in sheep's clothing. It is very friendly, and it can smile at you without a caution. Poverty does not respect your status, your stature and your age. It only gives credit to those who have purified their mind. A corrupt mind is a poor mind.

Poverty is very weird simply because it is unusual right from creation. Its power is beyond imagination and it is capable of ruining the whole system of human capabilities. Poverty does not joke with life but it can joke with your life.

Poverty was created by the human mind and since its creation, it has never rested. It sometimes roar like a lion and soar like an eagle just to stalk its prey. It comes without greetings and even decides to stay without your permission. It has no rule to obey and no pattern to follow. It comes with authority and power and very influencing in the minds of the poor. It is a deadly virus without mincing words. Everything about poverty is below ordinary and above headache.

Poverty began having effects on the body, the blood and the soul. The body became weak, pale and deficient. The blood had symptoms of diseases and the soul rejoice not. Everything about man was linked to poverty and it became the order of the day. Man suffered just because of poverty. Nature offered prosperity and man disobeyed nature's prospect. Things went from green to grey, good to bad, sweet to sour, extraordinary to ordinary and prosperity to poverty.

Things were in dismay due to man's illegal conscience. Man became his own enemy. The mind became corrupt, the body became sick and the heart dwelled in darkness. Perfections became imperfections and freedom became the thing of the past. The greatest enemy of man is not him but what is planted in him.

Every man was a replica of God on Earth. The image of God was pure and without fault. There was no poverty attached to God's image but man decided to alter such an imaginable state. Man was having dominion over the Earth. He was a king, he was a master, he was a boss, he was a leader, he was a controller and he was the god of the Earth. Everything submitted to his power until he sold his conscience to the opposition of his position. Opposition took over the Earth and injected what is known as poverty. Opposition is just like fire extinguisher. It is always there to quench fire and it has the ability to destroy. Man has encountered series of oppositions since the rise of poverty. There is no prosperity without position and there is no poverty without opposition.

The world has now become a battle field. War everywhere! The strong is battling, the wise is battling, the fool is battling and the weak is also battling. Confusion everywhere and even the innocent is initiated to a fight without a cause. The rich is suffering from depression, affection and opposition. The poor is also suffering from possession, affection and position. It seems everyone is poor and the mind is simply confused. Show me a confused man and I will show you a poor man.

TYPES OF POVERTY

CHAPTER 2

There are series of questions about poverty. Most people will ask, who is a poor man? What are the factors to be considered when subjecting poverty? Is a person poor because of no money, no car, no shelter and no food? What has poverty got to do with happiness, affection and position? How do you calculate a poor man's mind?

Poverty has levels no matter the subject under discussion. It might be generational, situational, relative and absolute. It can also be linked to economic, bodily, cultural, spiritual, political, societal and mental. No matter what type is being discussed, the mind is always there as a determining factor. The type of a man's poverty is determined by the type of mind he possess.

Generational Poverty

This is when poverty is transferred to individuals and families from generations before them. This type of poverty is chronic and most people are trapped in its causes. This is as a result of family having lived in poverty for at least two or three generations. Most people in this situation perceive they have no access to instruments that will assist them get out of it. Others think is a journey of no return. A man builds a house today, he dies tomorrow and the son takes over. He accommodates wealth, he dies tomorrow and the daughter takes over. He builds an empire, he dies tomorrow and the family takes over. He builds nothing and the family takes over nothing. Whatever a man builds on earth will surely be inherited.

It is the responsibility of every man to create and develop new things for the future generations. Humans have to create things of substance, things of hope, things of joy, things of abundance and things for the future. Things of value and not of loss. No creation, no inheritance. The Earth was created for our future and the legacy of every man is the life he lived for others and for the future.

Humans should have the capacity to build riches be it infrastructures, goods, money, goodwill, support etc. for the future. The future generation will live because we lived and for that reason we are mandated to live because they live. Man should not think about only today but tomorrow because that is the true purpose of creation. We owe our generations to come peace, love, kind, just, joy and sound mind. It will be the greatest error for them to come and meet agony. It is our duty to inculcate or plant right in our children.

Every man came to meet poverty after the first generation. The poor saw no wealth and the rich saw no peace. Man inherited disaster and disaster inherited man. Our great forefathers sowed the seed of poverty and it was automatically inherited by humans. Some destroyed the rich land nature offered and traded them for no good returns. They left high debt to be settle by the future generation. Others decided to infuse the whole system with corruption.

Wickedness has become the norm and we are too selfish to think right. Every thought of man is self-centered and family-centered. We now live in a world whereby it is very difficult for people to think of others and others to think of people.

Leaders of various organizations, institutions, clubs, families, societies and nations have become subordinated to selfishness. Their way of thinking is how they can become great and not how the future generation can be great. It is the duty of every head to protect properties and ensure equitable distribution of resources. Every leader is mandated to ensure corrupt free and flexible environment and so is the responsibility of every individual to invest positively in such environment. The environment within which we live has the tendency to affect our joy, glory, wealth and mind. Sound environment gives birth to sound mind and sound mind automatically gives birth to sound generation. When you tamper with any of these, you tamper with the generation yet unborn.

Situational Poverty

Situational poverty is a period whereby a person becomes poor because of sudden event. Situational poverty can be caused by so many factors such as; a turn in the national economy, natural disaster, sickness, loss of job, divorce or death of family head. Individuals fall below the poverty line because of sudden crises. A man in situation is poverty situated.

Sometimes, a lot of issues happen which beat human understanding. Things we do not expect come our way making life difficult and unbearable. Most people are trapped in such situation and they become handicapped. People in such situation find it difficult to get out of the mess and begin to start a blame game. Some even accept the situation and makes it situated. A person who finds reasons to point fingers is the person who has reasons to be poor.

Many of us go through disasters and instead of seeking solutions, we rather think of the problem. Things don't just happen, for every mishap there is a cause. You do not have to waste time on the cause but rather the remedy. A man can waste many years to think of why he is poor and not why he should be rich. Thinking about poverty is not the same as finding your way out. Situation in line with wrong attitude is a disaster. In this modern age of power, prosperity, fame, poverty and agony, you are what you are based on the decision you make. The way you respond to situations is the way situations respond to you.

Disaster falls on man and man decides to embrace it with negativity which is totally unacceptable. Humans are bound to encounter problems in this age of dynamics. In fact, there are unexpected crises that can beat human imagination but when this happens, it is our sole responsibility to count it void and prove beyond all reasonable doubt that indeed we define problems, problems do not define us.

You can only be in charge when you decode the power of positivity in every ugly situation. Every situation which is ugly is poor in nature. People will despise you because of uncomfortable events around you and that is one of the greatest sign of poverty. If people cannot accept you because of sudden crises, then you have no reasons to accept such crises. Break through and come out from the mess, for you are bigger than poverty. Allow poverty to be your slave and simply control it with a right attitude.

Most of the situational poverty that arises is from generational poverty. In other words, generational poverty breeds situational poverty. Some attitudes of our forefathers can surprise us with unexpected crises. Crises of which we have no control of excerpt the control of our mind. The way we respond to such crises will determine the state of our situation. Situational poverty has one major purpose and that purpose is to saturate the minds of those who are not well situated. Humans cannot stop certain poverty from happening but can stop certain poverty from ruling.

Let us look at this illustration; a person suffering from HIV/AIDs is in bondage but the same person suffering from such sickness might live longer than someone without any serious sickness. This is because the HIV/AIDs patient might be living under the medication of positivity which is the right attitude for such ailment and vice versa. Thus, the person without HIV/AIDs might also be living under the medication of negativity which is the wrong attitude which can cause death. You can cure ailments when there are positive wings to soar.

If situations occur, do not worry. Put up a smile and tell yourself, this is the opportunity to know some preferences and swing to the positives. If a person does not hunger, how can such a person tell when he or she is satisfied? It only takes lack to be able to know likes. Situational poverty is not a permanent poverty. It only becomes permanent when the person affected gives more room for accommodation. Whoever triumph over situations, triumph over poverty.

Natural disaster, sickness, divorce, loss of family, etc. can affect the wealth of a person. Some of these situations can even trigger the mental capabilities of affected persons. Most people do not even reason well when they go through these problems. They allow it to steal their minds and render it useless. We can overcome situational poverty by reducing accommodation and we can reduce accommodation by reducing negativity and we can reduce negativity by increasing positivity.

Relative Poverty

Relative poverty is the condition in which people lack the minimum amount of wealth needed in order to maintain the average standard of living in the society in which they live. What is poor in Ghana might not be the same as what is poor in Nigeria. It is the Inability to reach a minimum accepted standard of living in a particular society. The way Albert perceive poverty might be different from another person's sight for poverty. They might see things differently but what is poor is poor.

Poverty is relatively a mystery. A man who is rich might be considered poor based on the criteria under discussion. Are you poor because of no money, no cars, no house, no cloth, no family, no wisdom, no happiness, no fame or no respect? There are some men who are wealthy but no happiness and others are happy but no wealth. Whichever way, there is still some sort of poverty associated with them. Cruising the best of cars does not guarantee success and walking bare footed is not glorious either. It takes a man with wisdom to manipulate poverty.

We need wisdom to control money. We need wisdom to control marriage. We need wisdom to control happiness. In fact, we need wisdom to control almost everything. Wisdom is wealth and a man without wisdom is a man without wealth. You can only build a house when you think of it. There is no man on earth who can claim he personally built a house without the thought of building. The more your thinking works towards the right direction, the more you achieve success. Having the ability, strength or knowledge needed to gain prospects require wisdom.

Africa is rich in terms of natural resources and Europe is rich in terms of infrastructures. The opposite is true for that of poverty. Countries poor in gold and diamond might be rich in infrastructures. This simply means that your state of poverty might be someone's state of prosperity.

Relatively, poverty is measured by our way of thinking and our way of vision. The way we perceive things is the way we perceive poverty. There are some men who do not have physical cars to boast of but their minds have already driven the best of cars. Such people hardly envy others because they carry the mind of prosperity.

Absolute Poverty

It is a condition where humans do not have the minimum amount of wealth needed to meet the minimum requirements for one or more basic living needs over an extended period of time. These include things like food, health, sanitation facilities, safe drinking water, shelter, education, information, etc. This is a condition characterized by severe deprivation of basic human needs. It depends not only on income but also on access to services. These measurements vary widely depending on where a person lives for several reasons. For instance, a person living in a certain geographical area is likely to be paid less for rendering a service than a person living in a different geographical area.

Living in absolute poverty is disastrous and it can endanger one's life. There are a variety of factors that may reinforce or instigate the existence of absolute poverty, such as weak institutions and low level of growth among others. A person in this situation is more or less handicapped as far as poverty is concerned. Such a person is poor from all angles and there is no absolute way of exemption. A man in need is poor indeed.

There are some people on earth who are trapped in all areas of poverty no matter the classification. They are poor in everything and rich in nothing. In fact, their poverty is extremely poor. There is no prosperity in their dictionaries and no breakthrough in their mindset. Poverty is just the order and the rule of such people.

Most at times, people suffering from absolute poverty need a savior. Someone who can help them breakthrough. Such poverty can retard their mentality and even steal their power of reasoning. It takes people with power and resources to assist victims of absolute poverty. A man who lacks basic necessity such as food needs a source in order to be fed and a man who lacks shelter also needs a source in order to sleep. Every man needs a man but a poor man needs more than just a man.

All the things on earth are for some reasons. The plant, the sea, the land, the gold, the minerals, the knowledge, the wisdom, etc. are all there for a reason. Human race is now a competition and for that matter, a man without a competitor is an invincible man. People compete for power, gold, wealth and knowledge but can never compete for wisdom. Wisdom is the principal thing and a factor which determines how wealth could be obtained. No matter how absolutely poor you are, you can always pave through with wisdom.

Economic Poverty

An economy is an area of production, distribution, trade and consumption of goods and services by different agents. Economic analysis often progresses through deductive processes, much like mathematical logic, where the implications of specific human activities are considered in a "means-ends" framework. Humans want to run from economic hardship but they are not even prepared to meet an overflow.

Economics focuses on the behavior and interactions of economic agents and how economies work. Economics can generally be broken down into macroeconomics, which concentrates on the behavior of the aggregate economy, and microeconomics, which focuses on individual consumers. Macroeconomics analyzes the entire economy and issues affecting it, including unemployment of resources, inflation, economic growth, and the public policies that address these issues.

The health of the entire economy automatically affect the individuals. If the economy is sick, it means the people living in such economy are also sick. The same way applies to poor economy. When the economy is in poverty, it is the people who live in it that suffers. Poor economy is a poor individual and a poor individual is a poor economy.

Imagine an economy with poor policy makers, what will happen to such economy? The people are likely to be doomed under such policies and they have no option than to enjoy the fruit of poverty. Most of the weak economic countries simply lack the ability to employ natural resources for the appropriate use. There are still natural resources which have not been discovered and those that have been discovered are still not fully discovered. You can only discover when you have discovered.

There are many natural resources with different purposes such as gold, diamond, silver, iron, copper, minerals, land, soil, rocks, vegetation, wind, water, fossil fuel, etc. but how to utilize these resources is a major problem to most countries. The use of resources are counted because of discounted discoveries.

A sustained increase in the general level of prices for goods and services in a country can be emphatically linked to poor policies by policy makers. Weak policies are likely to retard economic growth and development. Therefore the rich becomes richer and the poor becomes poorer. High productivity is the utmost key to economic freedom and must never be jeopardized.

Microeconomics analyzes the behavior of individuals and firms in making decisions regarding the allocation of scarce resources and the interactions among these individuals and firms. It deals with the choices and decisions that have to be made in order to manage scarce resources available to us. When resources are scarce, it is our choices that can scrap the power of scarcity.

The world is rising faster, and the economy is sinking at an increasing rate. It takes outstanding individual behavioral to fight such canker. The use of resources is not fairly distributed and individuals who are to make right decisions are making wrong decisions. The economy is in shambles because of bad decisions made by broad individuals. The number of individuals who misuse resources are more than those who economize it. Economic poverty is as a result of waste or unnecessary expenses relating to firm, domestic or household affairs.

Bodily Poverty

Body can take many definitions such as the material organized substance of an animal, whether living or dead, as distinguished from the spirit or the trunk, or main part of a person as distinguished from the limbs and head. Another definition can be taken as a number of individuals spoken of collectively, usually as united by some common tie, or as organized for some purpose.

The body we want to talk about is a person and not just a unit. Bodily poverty is the lack of properly physical health and access to health living conditions. We have to look at both the outer and the inner body of human. The inner body of a person regulates the activities of the outer body. It is very necessary to take much care of the body and to keep it safe. Your attitude towards your body is the true reflection of you.

What you eat, what you drink and how you think are the three main factors that contribute to a poor body. You cannot eat poor food and get rich body. You cannot take in poor drinks and get rich body and you cannot think wrongly and expect to have a correct body. Whatever positive or negative thing you consume will automatically have an impact on the body.

Most people do not eat, drink and think right but they are expecting a healthy body. Food is one of the most essential nutrient the body needs. Man have to eat rich in order to have a rich body. To eat rich does not mean to spend thousands of dollars on food or to eat a bag of rice. To eat rich simply means to balance what you eat even if it is small at the right time and at the right place.

Another factor which can tamper with our body is what we drink. You cannot drink poison and live. People normally say water is life yes it is true but water is life when it is safe. The safer the water, the richer your body and vice versa. If truly you are what you eat, then there is no reasonable doubt that you are what you drink. A body without water is not a healthy body.

Finally, what you think is the major factor that contribute to our poor body. There is power in our thoughts. Our thinking pattern has the power to control what we eat and what we drink. After all the good food and the best drinks, our body needs sound mind in order to utilize what we eat and what we drink. A person who is emotionally enslaved lacks the ability to taste good and a poor thought has the power to neutralize nutrients.

Cultural Poverty

Culture refers to the characteristics and knowledge of a particular group of people, encompassing language, religion, cuisine, social habits, music, arts and other manifestations of human intellectual achievement regarded collectively. We have rich culture and poor culture but the emphasis here is on poor culture. This type of poverty is the result of certain cultural values passed down through generations. In most cases, the result of people's values or cultural norms is poverty.

Most people who are poor have different cultural values and different perspective altogether. Such people are like refugees in their own country, convinced that the existing institutions do not serve their interests. The environment in which a child grows has the tendency to depict the behavior of the child in the near future. What grows a child is what keeps a man.

It is very outrageous to know how some cultural values demean prosperity. In some part of the world, it is even a taboo to own a car at a certain age. A taboo to acquire some skills. A taboo to build mansions and even a taboo to acquire certain level of knowledge. Some perceive too much of money to be evil and others think you cannot break the norm as far as their values are concerned. A person from such culture can only breakthrough when he or she meets different culture with different perspective. A primitive culture is a poor culture.

The type of culture we have these days is a different thing altogether. We now have a culture of competition, culture of corruption, culture of selfishness, culture of hatred, culture of envy, culture of jealousy, culture of wickedness and the culture of destruction. These are the very cultures we pride ourselves in and the very cultures we desire. Our way of life is now more of negativity and less of positivity. I was told culture is love but later discovered it was a culture of word and not a culture of action.

Children are learning from their parents how to be selfish. Students from the basic, secondary and university level are learning from their teachers how to be corrupt. Employees are learning from their fellow workers how to compete. Individuals are learning from their community how to initiate conflict and nations are learning from their fellow nations how to go for war. How do we prosper with these kind of cultures?

A child who watches his father beats his mother will automatically learn the art and even do it better when he marries. Nations are fighting among nations, communities are fighting other communities, individuals are fighting other individuals and here we are, expecting progress. The outcome of these cultures only transforms people into refugees and nothing else. Anything which has the tendency to bring out the beast has the power to kill the beast. There is only one true culture which can eradicate all these canker and that is the culture of love. No matter our background, belief, religion or our differences, there is always a way out and that way is the way of love.

Spiritual Poverty

Spirit is the intelligent, immaterial and immortal part of man that makes any supernatural being, good or bad. Thus the soul, in distinction from the body in which it resides. A lot of people do not understand spiritual things and those who understand sometimes do not believe spirituality. It is one thing to understand something and it is another thing to believe it. Whatever a man believes in has the power to behold him.

Man was created in the image of God and that makes every man a supernatural being. Spiritual things are hidden because it needs discovery to make it active. We are all gods on earth because we represent the Supreme God but the issue here is, not all of us enjoy such privilege due to lack of discovery. We do not know whom we are and we have embraced only the humanity part ignoring the spirituality of the earthly race. Most humans are only human beings because they have not discovered their spiritual being. They are spiritually poor and carnally minded. Everything relating to the spirit is not active in spiritually poor humans. Being spiritually rich is not that difficult but it is difficult for the mind that do not perceive them.

Any man who want to be spiritually rich will have to follow these steps;

1. Believe in the existence of a Supreme God.
2. Believe that you are the exact replica of God on earth.
3. Have a self-retrospection.
4. Meditate day and night.
5. Feed your soul with spiritual food.
6. Enrich your spirit.

Someone may ask, who is the Supreme God? The Supreme God is the alpha and the omega, the beginning and the end creator of the universe. It is only God who knows the beginning and how things are going to end. The moment you recognize the existence of God is the very moment you will recognize your true identity. Believing God align your spirit with His which empowers you to triumph. Our true self can only be found in God which is due to begin and due to end. If you cannot find God, find you and if you cannot find you, find God.

To be the replica of God is to represent God. God is a spirit and whoever worship God must worship Him in truth and in spirit and it is therefore also paramount to represent Him in the physical. The richness of God is deposited in you and you have every right to tap into it. The choice is yours. Your living should reflect God. When people cannot find God, let them come to you and when they come to you, let them see God because you are the exact replica of God on earth.

To have some retrospection is very necessary as far as spiritual poverty is concerned. Examining things will help correct certain things in the future. You only have to recognize your mistakes and build a better future. Your spirit being should be able to get rid of any diabolic waste within the spirit. There are a lot of things that hunts our spirit being and it must be detoxify.

To meditate simply means to consider, ponder, weigh, revolve, study and to plan. We meditate a design when we are looking out or waiting for the means of its accomplishment. You have to keep the mind in a state of contemplation and dwell seriously in thought. To meditate is to speak to the inner spirit, to speak to the inner spirit is to discover the inner spirit, to discover the inner spirit is to discover God and to discover God is to represent God.

There is no human on earth who can proudly claim that he or she has not eaten for a decade. The body needs food to survive, the same way the spirit also needs food to survive. We all know the type of food we take in as far as the body is concerned but it is absolutely different when it comes to the spirit. The type of food the spirit needs the most is love, kindness, forgiveness and righteousness. These type of food empowers the spirit into richness and directly connect you to the Supreme God. A man who protects his body and ignores his soul is just like a machine without a proper engine.

To enrich your spirit depends solely on the type of spirit you want to build. What type of spirit are you building? Is it the Holy Spirit or the evil spirit? Be wary of the kind of spirit you want to build because the type of spirit you initiate has the power to make or destroy you. The Holy Spirit is the only true spirit to curb spiritual poverty. What then is the Holy Spirit? The Holy Spirit here means to be spiritually whole or sound which is free from sinful affections. There is no way a man can be spiritually rich without having the Holy Spirit.

Political Poverty

Politics is that part of ethics which has to do with the regulation and government of a nation and how political party is managed. It is obvious that a nation's poverty rate is sometimes quantified by how authorities disburse national resources amongst the people. Politics can affect people in their personal lives and may generate frustration and chaos if not properly managed. Politics benefit from everywhere but not everywhere benefit from politics.

The true purpose of politics is the preservation of safety, peace and prosperity but it is rather opposite for political poverty. Politics seek to protect the rights of citizens, improve their morals and augment resources. Thus, the true nature of politics. The world of today has embraced the false nature of politics which is political trickery and this is the very root of political poverty.

The conducts and contests of parties with reference to political measures is a different game altogether. It is a game of trust, promises, fraud and hopes. The conducts and contests through which the game is carried out will determine the ultimate success or failure of public affairs. Politics of trust and hope is a necessity to national development and without these, there is no politics.

The administration of public affairs is likely to be in shambles when the art of fraud and promises are mastered. The foundation of most political parties are built on deception, fraud and promises. Politics of deception, politics of fraud and politics of promises have become the order of the day and these acts have negative effect on the administration of public affairs. Poverty is likely to reign when promises are broken and resources are channeled into few pockets. The promises of political parties these days can build another planet if they are to go according to what they promised.

Societal Poverty

Society has always been a point of relationship where individuals connect, participate and partner with other individuals in order to maximize a certain level of desire. Persons of all caliber including friends and acquaintances are united together by a common bond of familiarities. The prime purpose of every society is to come together in order to help each other but the society we see today came to compete, steal and destroy.

When the rich gives to the society, it automatically reduces poverty. Societal poverty has always been as a result of selfishness on the part of individuals. People do not find the need to help other people and all what they are interested in is themselves. Societal poverty depicts suicidal individuals within the society.

The foundation of every rich society is giving. The opposite is true for poor society. When you check the status of great societies, you will realize that each and every one has something to give. Some might give materially whiles others give mentally and that sum up to their greatness. The power of giving is a supernatural rule of nature which propels prosperity and repels poverty.

Mental Poverty

The mind is just like the central processing unit of a computer and without the processing unit, there is no computer. The same applies to humans. Which simply means human beings cannot live without the mind and it is the very part that controls every part of the body. In other words, the mind is the center of a man's life. The strength of every man is in his mind and it is very difficult to live an extraordinary life with an ordinary mindset.

It can be deduced from the root that poverty is the state of desperation, inability to possess possessions and the total confusion in the mind of those who lack the ability to change situations.

When a man is poor, check his actions. It is good to strive and not to sound desperate because desperation is a clear sign of despair and hopelessness. A hungry man will automatically chase food which comes at no cost and most at times, we chase what we lack in real life without thinking of the repercussions. Show me a desperate man and I will show you a poor man.

Before you can change poverty, you should be able to think of the possible solutions first. Stability of mind is needed to pave through hard times. The mind is a powerful tool which can never be underestimated or neither can it be tampered in any way. You cannot think wrongly and live a righteous life and the same way you cannot think poorly and live a wealthy life. It is very necessary to think rich in order to have a fruitful life because it is very difficult to live such a life without thinking right. There is power in positivity to overcome negativity to produce prosperity.

One of the greatest threat in this generation is how we think and how we effect change with our thinking capabilities. There is something in the mind that helps us to nullify every complications and that kind of mindset can only be decoded through this process;

1. Define the problem
2. Analyze the problem
3. Find solution to the problem
4. Solve the problem

Without the definition of the problem, there is no problem. Every problem has to be defined in order to be revealed. Defining the problem is just how you see, perceive and think. If your poverty is material poverty, so it is and again if your poverty is mental poverty, so it is. It is well defined by whomever it is perceived. In this era of tribulations, you have to define problems and do not allow problems to define you.

Analyze the problem after the definition to examine in such a manner as to ascertain some vital elements. Break down the problem into components or essential features which can be subjected to psychoanalytic treatment and find possible solutions to the problem.

You should be able to finally solve the problem with the possible solutions at hand and if you cannot solve the problem, just go back to the definition and define the problem again. This time, let your definition be like this, "there is no problem."

Let us analyze a case study concerning a mad man. A mad man can walk 5000km and more without getting tired. A mad man can eat anything without being sick. In fact, a mad man is only mad in the sight of those who perceived him to be mad and just that his mind has been programmed to walk and to eat anything.

All what we have to learn from this case study is that, in life everything is possible as far as the mind is concerned. Let your mind be mad for prosperity, let your mind be mad for victory, let your mind be mad for greatness, let your mind be mad for knowledge and let your mind be mad for wisdom. You only have to program your mind in that direction in order to fight and triumph over poverty. Whenever you are overwhelmed by certain level of poverty which is beyond your capabilities, just relax and say to yourself, in this era of tribulations, we define problems, problems do not define us.

AGENTS OF POVERTY

CHAPTER 3

An agent is the one who exerts power to produce an effect or has the power to act and cause change in the life of individuals. Agents of poverty are the people who have been entrusted with power and authority to cause positive change but the change they effect always result in poverty. These agents include politicians, religious leaders, teachers, students, parents, employees, employers, wrong people and individuals.

There is no undisputed fact that these people are doing tremendously well but the emphasis is on the negativity as far as poverty is concerned. We all know the benefits we get from politicians, teachers, students, parents, employees, employers and individuals. We also know their impact, the blessings, the advice and the progress they have to offer but we will look at the crimes or sins they commit in order to bring poverty.

Politicians

Politicians are well versed or experienced in the science of government and they are very devoted to their own advancement in public office. It is a great privilege to be a politician because at some point in life, life depends on it. The decisions they take affect people, nations and the world at large. A politician contesting for power is different from a politician in government.

One of the greatest crime or sin of politicians is selfishness. All politicians who contributed positively were selfless and that is the exact secret to greatness. You can only rate a poor country when you have rated the attitudes and actions of its politicians.

Politicians who are selfish will never think about the future. All what they are interested in is today. They will take decisions for only today, borrow money for today, buildings for today, sign contracts for today and everything is about today. They prioritize their dealings to suit what they want today and not what the people will need tomorrow. Majority of today's politicians are politicians of today not because of today but because of selfishness.

A politician of vision might rise to release hope into the future but a politician of mission is already established in the mind of innocent individuals. Poor nations are very familiar with politicians who came to steal and to destroy and not politicians who give hope and a future.

The only difference between a vision and a mission politician is that, a mission politician's vision is to miss the future and a vision politician's mission is to mold the future. The future is not in the curriculum of a mission politician. His mission is to steal and leads the people astray. They create a comfortable life today and inject poverty into the future.

Religious Leaders

Religion is an act by which humans depict their recognition of the existence of a God or gods having dominion over their life, to whom obedience, service and honor are due. There are a lot of religious leaders in the world which include pastors, prophets, imams, priests, etc. and these leaders are responsible to bring positive changes to the people. A religious leader who only represent God or gods and not the people is not a reliable leader.

A true religious leader is the one who represent God and the people. Most religious leaders represent God and do not want to represent the people. They have God at heart but do not have the people at heart. They know the will of God but do not want to do His will. They know the problems of the people but do not want to solve their problems. What they do is to add more problems. The God we serve is a solution and selfless God but most religious leaders who represent God are only solicitors of selfishness.

Religious leaders are the very people who have the capacity to change the world simply because the largest population is of religion. Almost each and every one belongs to a particular religion and that makes it the biggest media of transformation but it is rather unfortunate that the biggest media of transformation has now become the biggest media of confusion and stagnation.

God has the power to break protocol and to cause change and because of that people always want to take advantage of such powers. Most religious leaders misuse such powers to favor only them. Religious leaders misuse the powers of God to the extent that the ordinary individuals see such powers as a way of escaping hard work.

Farmers, traders, carpenters, plumbers, electricians, repairers, etc. do not want to work hard anymore because there is a religious leader somewhere who has promised them breakthrough to be better than whom they are. It is good to be better than what you are but it is not really good to be better than what you are not.

Hard work is one of the keys to prosperity and every nation's success is as a result of its hard working citizens. Instead of seeking strength to enable us work hard, we rather seek powers that can work for us and this is exactly what has led many nations to be poor. Working hard is not about burning your intestines alive, it is all about working right without giving up. Most religious leaders place emphasis on what God can do for us and not what we can do for ourselves. God is always willing to provide and He is faithful to do exceedingly and abundantly. His Grace is sufficient to meet our needs but what one has to know is that, it is our responsibility to tap into the Grace with focus and determination. Religious leaders are agents of change and that change must be a change of mind, a change of health, a change of laziness, a change to prosperity and not a change to poverty.

Teachers

A teacher is someone who instructs others but the emphasis will be placed on formal education or classroom teachings. A teacher is the most essential resource in the educative process because they are the people assigned with the task of imparting positive changes in children and the young adults. A teacher is always an instrument of change in the life of the pupils assigned to teach and every action of such a teacher is tantamount to creation or destruction.

The best time to fight poverty in the life of an individual is during infancy. Children grow with what they see, what they hear and what they learn. Imagine a child growing in the midst of armed robbers or a child growing in the midst of drug addicts. There is a very high tendency of such a child becoming an armed robber or drug addict. The same way applies to class management. The way a teacher manages a class is very paramount in his or her career. Poor classroom management is likely to breed poor future leaders and teachers who have to effect desirable changes are causing miserable changes in the life of innocent pupils.

A teacher should always assure the pupils that they can make it. He should be the center of motivation and speak positive words over the pupils. His or her words should be words of encouragement and prosperity. The future must be created with positive words, thinking and actions in the classroom. But most teachers rather seek to demean the pupils based on their thinking, actions and performance which in one way or the other affect progress.

The teacher should always have it in mind that he or she is dealing with pupils of different capabilities and skills. There is nothing like poor or weak in the educative process simply because the poor or the weak also have a strategic way of soaring higher in life. It is the responsibility of the teacher to discover and nurture hidden talents or skills of pupils. Sometimes, pupils who find it difficult to learn or study new things are the very pupils who have the power to reject old things. It is the duty of the teacher to help the child in a strategic manner in order not to be left out in the educative process. That is why a teacher has to be psychologically trained in order to help the pupils.

Students

Students who attend a school and seek knowledge from professional teachers or from books are obliged to study or examine things in an attentive and systematic manner. Most students learn for learning sake and not for purpose sake. Some students learn because they just have to learn and not because they have a justifiable reason to change the future.

Majority of students do not even know why they go to school and that is one of the major crimes of most students. Education has become normal for students and the only reason they perceive is to pass their exams and get a government or better job in the society. It is good to pass your exams and get appointment but that should not be the most fundamental reason for going to school. The most fundamental reason for going to school is to discover your hidden talents and how the school is going to help you maximize such talents in order to help your society and the world at large.

It is very unfortunate that most poor nations produce paper graduates and not purpose graduates. Students can use three or four years studying why it is good to farm and not why it is good to be a farmer. How can a nation produce such graduates and be expecting to flourish or prosper? They know the benefits we get from farming but when you tell an agricultural graduate or officer to even hold a cutlass, disaster! Students in most developing countries are trained to know and not trained to do and this lack is as a result of negligence on the part of both students and teachers.

Parents

Parent refers to a father or a mother who begets or brings forth an offspring. Creation has been done from one generation to another through parents and that automatically makes them key stakeholders in the creation process. They have unique powers as far as child upbringing is concerned. Parents are the very people who can add or subtract from the future of their children.

There are a lot of crimes parents commit towards their children but the most popular ones are;

1. Lack of proper foundation
2. Use of wrong words
3. Inability to invest in the general wellbeing of the children

It is the duty of every parent to establish proper or a concrete foundation for the children and the generation yet unborn. Most parents are very lazy and they see no reason to work hard just because of their children but the truth is that, it is mandated for each and every parent to work hard not because of today's hunger but the hunger of tomorrow. They think it is okay to be satisfied today and not tomorrow and because of that all their effort is channeled against tomorrow. The foundation you will create will determine the future building and building which lacks the ability to accommodate riches is as a result of poor foundation.

Another most dangerous crime they commit is the wrong use of words. Words are very powerful and not only powerful but extremely powerful to create and to destroy. In fact, words have the tenacity to bring things to life and it is the sole responsibility of every parent to protect their children physically, spiritually and mentally with their words. Parents rain curses instead of blessings and demean their children just because of their mistakes and this can cause emotional tremor and psychological poverty. The type of words you declare over your children will determine the type of world they will build with your words.

Inability to invest in the general wellbeing of the children is a wicked act towards children. Most parents can invest in their children but they will deliberately refuse. They will rather buy themselves new clothe, jewelries and enjoy the best food in the best restaurant. The meat alone on top of their food can buy books for their ward. Parents of this age desire luxury lifestyle than healthy lifestyle in the expense of their children's wealthy lifestyle.

Employees

An employee is the one who has been employed by another to carry out certain responsibilities at a workplace. The only serious crime of most employees is their negative attitude towards work and such attitude determines the progress of a nation. Productivity has a way of reducing poverty and the only way to reduce poverty through productivity is to produce more. The negativities of employees which hinder productivity includes;

1. Laziness
2. Time
3. Competition
4. Selfishness

Employees from different background come together to exhibit different behaviors and the most popular among such behaviors is laziness. Laziness is just a mental catastrophe of both physical and spiritual manifestations. Which simply means that when the mind is lazy, it automatically affects every area of your life. In fact, laziness is now a hobby for most employees and it is being considered as normal. Some employees who are not lazy too have intentionally cultivated laziness simply because they do not see the reason why they should work extra hard for someone's organization to prosper.

The second negativity of employees which hinder productivity is time. Time has a way of changing situations but not when that situation is under the influence of man. When an employee's attitude to time is poor, it is the organization and the nation that suffers the most and that is a clear sign of wickedness on the part of such employees. If you do not respect time, time will never respect you and if you do not work within time, time will never work with you, this is what breeds poor individuals, organizations and nations.

Competition is a good tool for productivity and creativity but not when the contest is opposed with struggle. The basis on which people compete these days is not ideal for advancement. Employees compete with the motive of bringing another person down and not the motive of increasing production. When your competition is not yielding any positive result, then it means your competition is not a healthy one. Employees compete in order to increase production and not to decrease production so whenever there is decrease in production then it means that the positive motive of competition has been tampered with.

Selfishness is one of the most dangerous deoxyribonucleic acid (DNA) of humans and it takes only a selfless mind to neutralize such canker. Employees only care for themselves at the expense of others and organizations. The only thing they think of is their salary and other benefits from workplaces without taking into consideration how and where the money will come from. They are too passionate to receive their wages and not that passionate to increases production and services. Employees are the key stakeholders in national development and majority of the powers are invested in them to make a change but the only change most of them make is the change to suit themselves.

Employers

Employer is the one who employs another and takes charge over their responsibilities at the workplace. He or she is more or less the manager or the director in that organization and responsible for the general wellbeing of employees. Employers are very unique when it comes to job creation. They have the technical skills for creating jobs but the ability to manage jobs is what most of them lack. There are a lot of crimes they commit during the process and this includes;

1. Evasion of tax
2. Unhealthy competition
3. Wrong recruitment
4. Emotional instability

Tax is a duty or levy laid upon specific things or organizations or the sum which fuels the system of government. Tax is needed to develop a nation and that is why it is very necessary to pay tax. Most employers are very wicked and dishonest. They always evade tax and that is one of the reasons why most countries are not developed. When the task of employments are not taxed, the tag of poverty is rather taxed.

Employers or organizations that engage in unhealthy competition do more harm than good. You cannot compete with your competitor by doing the same thing he or she is doing. It does not bring much result and it kills uniqueness. You compete with the idea of being unique and when uniqueness set in, people's needs are satisfied.

Another crime employers commit is wrong recruitment. They choose the wrong caliber of personnel to fit into the organization and this has a strong negative effect on production of goods and services. You cannot employ someone who is not capable just because he or she is your friend or family member. Employing someone who does not qualify for the job will only ends up disqualifying your products or services.

Emotional instability has a powerful way of disturbing or hindering productivity and this has been a major challenge for most employers. Employers are sometimes easily moved by both internal and external environment within which they find themselves. It is good to have a sensational mind or soul but it is not good to apply such senses on things which does not make sense. It will only destroy your inner peace and nothing else. When employers are emotionally instable due to errors of employees or other factors, it might just create communication and relationship barriers which will affect work negatively to poverty.

Wrong People

Human beings are interrelated and it is absolutely not receptive to be isolated. We need each other to survive on earth, not after death. Everything we do on earth is connected and that makes it difficult to stand alone. You need someone to give birth to you and to take care of you before you mature. You still need people even after maturity to order your steps to greatness. That is why we all need people to survive on earth. Due to the fact that we need people has permitted wrong people in our lives. Most people are agents of poverty in our life because they desire to plant wrong instead of right.

The person you walk with has the tenacity to bring out the good or the bad in you. Life becomes easier when you are with the right people and very cumbersome when you are with the wrong people. The high rate of poverty these days are as a result of wrong connections. When you are located at the right place with the right people, your greatness is highly assured. Humans are responsible for solving each other's problem and if you cannot solve any problem, it means you are a problem to our problems.

A president, minister, priest, banker, chief executive officer, teacher, doctor, nurse, engineer, etc. are all likely to fail when they are connected or surrounded by wrong people. A president of vision will be elected into office and the next thing you will see is retrogression instead of progression. It will look as if the president is not doing well but the truth is he or she is capable but the people connected to him or her might's not be capable of doing what is right. It is just that the president is only connected to the wrong people at the right time and sometimes, these very people are capable of doing what is right but they simply desire to do what is wrong. The same applies to everyone. When the wrong people are connected to our life, it plough our right and automatically leads us to our plight.

Engaging in an illicit relationship is one of the dangerous ways to initiate a man into poverty. You should always try your best to stay holy when you are not married to avoid future discrepancies. Whatever affair you engage in today will affect your future so do not entertain people who will arouse your desire when the time is not right. Connect to people who will encourage you to study and work hard, people who will motivate you to do the right thing and not people who are only interested to send you to wrong places and other joints of no benefit. Do not rush to do otherwise when you are not married. Every person should be careful in choosing his or her spouse when it is time for marriage because the kind of person you connect to determines the kind of burden you will carry.

Individuals

Individuals possess different mind of their own that distinct them from all other beings. Each and every one has a peculiar traits, characters and personalities that empowers to uniqueness. You can never be someone else but you can try to be someone else. Which means you can learn the art of another and act like that person but you cannot be exactly like the person no matter how we try. We perceive others to be powerful just because we do not want to receive the power to be grateful.

One of the greatest crime of most individuals is the crime not to accept whom they are and what they were created for. We have become our own enemies and failures. Everything we are meant to be has never meant to be because we did not mean it to be. Have you ever ponder over these questions?

1. Who am I?
2. Where did I came from?
3. Why was I created?
4. Where am I going?
5. How will my final destination be like?
6. What can I do best that others cannot do?
7. What am I doing that everyone is doing?
8. What makes me fulfilled?
9. Do I add positive value to the world
10. How do I fight poverty through positive means

All these questions are positive questions to reflect your real self or discover your self-identity. You cannot fight poverty without being a thinker. You have to adopt clear strategies to withstand poverty affecting every area of your life and this is what most individuals have chosen to ignore. A lot of people look at what people can do and not what they can do. They look at what people are doing and not what they are doing. You need focus and determination to achieve an objective and the only way to be focus is to ignore what others are doing.

A lot of people have become burden on others simply because they have refused to accept realities in life. Most individuals are too lazy to the extent that they want others to fight poverty for them. In fact, they even want others to think for them. Poverty is a force and any individual who want to fight poverty also needs a force and not just any force but extraordinary force.

CHARACTERISTICS OF A POOR MAN

CHAPTER 4

Characteristic is a distinguishing trait or behavior, quality, or property whether positive or negative of a person or thing. Every man has a peculiar or distinctive characteristics that makes a person different from another. Most people might argue that one of the major characteristics of poor people is unemployment, and others might also attribute it to lack of opportunity. Such people will always have problems with the job market and justify reasons why people that live in a location with a limited number of good-paying jobs have far fewer opportunities.

Sometimes, the person you consider rich might be poor and the one you consider poor might be rich. The attitudes and behaviors of people determines how poor or rich they are. Your material wealth should reflect your attitudinal wealth and riches without the right attitude is witches. There are a lot of characteristics that tell how poor a man is and the following are some distinguishing traits;

1. Poor people blame, complain, criticize and condemn. One of the strategic ways to identify a poor person is to check the manner in which they complain, criticize and condemn other people who are doing well in all endeavors. A poor person will always give reasons why he or she is poor. In fact, he or she has people such as friends and families, governments and the entire nation to blame for being poor. Poor people really do have time to talk and discuss about other people's business. A poor man's desire is to discuss other people's business in the expense of his or her own business. They find it difficult to rate success and sometimes master the art of condemning true success. Leave the witch, demons, family, bosses, pastors and the government alone, rather learn from their mistakes and work on them to achieve your goals. The more the excuses and the criticism, the poorer you become.
2. They think money is more important than time. Money is a powerful tool and precious like gold. It is very important because it serves many purpose. Time is the period or the duration of one's life at which any definite event occurred. Time can be traded for money but money can never be traded for time. You can always get your Ghc1000 back, but you cannot get 1000 hours again. A poor man hardly respects time and sees no need to work within time. Time is money they say but money is not time. A man who is poor will use all his strength to chase money instead of working within time.
3. They want the cheapest way. When you come across a person who always want to receive something and not to give, you have simply met a poor person. A poor man is characterized by easy and free-ways. They want things which does not demand sacrifice and they are also interested in things of no

value but of substance. A man can only be considered cheap when his strength, time and mind are of small value.

4. They cannot control their greed. Greediness is the quality of having a keen interest or desire for anything. It is good to desire something but it is very bad and dangerous to desire anything. Greedy people will eventually lose everything because they chase everything. No matter your possession, when greed is your quality, then you are poor.
5. Poor people lack financial discipline. It does not matter how much you make from your job, if you are indiscipline with your finances you will find it difficult to progress.
6. They lack focus. People normally say that poor people do not have goals but the truth is everyone has a goal, what poor people do not have is focus to achieve their goals.
7. They make assumptions. An assumption is the act of supposing a thing without proof or unwarrantable claim. Poor people do not set realistic and attainable goals and because of that they always live in assumptions.
8. They give up easily. A person who do not have the strength to withstand pressure is poor in character. It takes strength to overpower obstacles and without it, man is vulnerable to failure.
9. They never learn new things. It is very appropriate to educate yourself on financial and other matters when the opportunity avails itself and you should always be willing to seek both formal and self-education. In fact, poor people are afraid of beginning and learning something new. They keep on doing the old things that hinder progress. The world keeps on changing and you can only adapt when you learn new things.
10. They engage in unnecessary competition. Poor people compete, rich people create. A man whose only motivation is competition is a poor man. No matter the level of your wealth, when your true desire is to compete, then your true character is poverty.

THE FORCE OF POVERTY

CHAPTER 5

Force is a coercion, violence or compulsory power exerted against will producing an effect, strength or energy of body or mind or any action between two bodies which changes their relative condition as to rest or motion. The forces of poverty are the strange powers that oppose progress, riches or wealth. An advancement or progress is needed in every area of our life because it is definitely the source of our joy. Riches do not consist in having more gold and silver, but having gold and silver which do not contradict peace of mind.

It can be deduced from the root that there is absolutely nothing good about poverty. All wrongs are poor and all rights are rich. Thus, all the wrong issues connected to your life are bad in nature and all the right issues connected to your life are good in nature for right is rich and wrong is poor. Every person desire prosperity but the question is; why are a lot of people poor? This question can only be answered when we take a critical look at what forces oppose prosperity. Have you ever asked yourself these questions?

1. Why do I find it difficult to learn?
2. Why do I find it difficult to focus?
3. Why do I find it difficult to persevere?
4. Why do I find it difficult to give?
5. Why do I find it difficult to help?
6. Why do I find it difficult to work hard?
7. Why do I find it easy to receive?
8. Why do I find it easy to play?
9. Why do I find it easy to relax?
10. Why do I find it easy to discuss about others?
11. Why do I find it easy to engage in unhealthy competition?
12. Why do I always want the cheapest way?
13. Why do I give up easily?
14. And why do I find it easy to do bad things instead of good?

All these questions have answers and their answers can be derived from what is known as the force of poverty. There are strange forces behind all these problems and such forces are not just mere forces but extraordinary forces. There is a general answer to all these questions and the answer is because the work of poverty is to prevent you from living a progressive life and it can only achieve that by releasing difficult forces.

When you find it difficult to study or learn, it is because the result of studying or learning is knowledge and knowledge in itself is a threat to poverty. So the best thing poverty can do is to release a strange force that will make you uncomfortable during studies so that you can forgo learning. When you learn you will get knowledge and when you get knowledge you will prosper so in order for you not to be knowledgeable, there is a force which is always there to make learning a headache.

The same applies to focus, perseverance, giving, help and working hard. There is a great reward in giving, help, persevering, focus and hard work. Poverty will not just watch you prosper with these powers. It has to release a force to demean your interest in giving, helping, perseverance, focusing and hard working. Laziness is one of the greatest force release by poverty to undermine hard work.

It is easy to do things which are bad because bad things do not add value to life and things which do not add value are the things which does not come with a cost and whatever thing which does not come with a cost is not difficult to do.

There is another force of poverty known as spiritual force. This type of force is connected to the spiritual realm but the end result will manifest in the physical realm. Most young people of this age are ignorant of these forces and because of their ignorance, they are hold captives. This type of force can attack anybody but its core mission is to attack the youth because they represent the current generation. Let us take a look at some of these forces which hinder youth development.

1. Money force
2. Marriage force
3. Sickness force

Money is a legal tender, which is used for the payment of goods and services and there are so many sources of money which include sales from products, daily wages, salary workers, and so on. There is a force that restrict and opposes money and marriage and there is another force that releases sickness. There are so many marriage problems which are as a result of spiritual forces. Your connection to certain idols may bring such problems in your life and if care is not taken, it will be transferred to generations yet unborn. A lot of individuals are suffering because of these forces and sometimes, they have no idea what actually force their money, marriage and sickness. You can only fight these forces by availing yourself to meditation and prayer and when it is done, you can pursue your desire.

THE YOUTH IN POVERTY

CHAPTER 6

The world is made up of the child, the youth and the old. The child represent the future generation, the youth represent the current generation and the old represent the end generation. All these generations are interconnected and one cannot live without the help of the other. The future generation depends on the current generation and the current generation depends on the end generation and vice versa. Poverty pays much attention on the youth than the child and the old simply because the child still has the power to change the future and the old has the power to go with the future but the youth determines the future.

Every decision of the youth will determine the success of the child yet unborn and the old yet to end and most at times both the child and the old depend on the youth for survival. That is why poverty is always interested in jeopardizing the youth. Majority of the youth in this age are ignorant of the fact that life is really not what it seems. The way they think, see and assume things are totally different from reality of life. Life is not assumptions, it is a reality. The following are some of the ways the youth perceive life;

1. Easy way to get money. Most youths are funny and ignorant of the fact that there is no shortcut or easy way to get money. You do not have to get money, you have to earn it. Money gotten without earning still needs some work in order to be maintained and that should tell you everything about money involves work. There is always a price to pay when you want to opt for easy way to get money and that price will only make life and money difficult.
2. All that glitters are gold. The youth normally think all that glitters are gold and this is one of the dangerous threat to the wellbeing of the current generation. It is not every riches you see is riches and not every success you see is success. There are a lot of deceptions connected to some luxuries and it takes a wise man to differentiate between the two. The youth who normally chase everything that glitters are likely to face grieves in the future.
3. They have all the time in the world. The youth perceive the world as theirs and because of that, they think all the time available is also theirs. They live carelessly without taking the repercussions into consideration and often commit mistakes that cannot be corrected.

The youth's perception about the world makes them easily fall into poverty. Their thinking pattern easily attract poverty without their concern. They can eradicate poverty by taking the right steps in life. The best time to use your energy is when you are young and your energy will only attract two things that is poverty or prosperity depending on the use. If you want to attract prosperity, use your energy wisely.

THE SACRIFICE

CHAPTER 7

Sacrifice has a broad meaning and the very meaning we want to take into consideration is the offering, destruction or surrender of anything valuable for the sake of obtaining something. There is a reward after every sacrifice but there is a greater reward after every good sacrifice. Which means that there is a bad sacrifice and a good sacrifice but the emphasis here is on good sacrifice. You have to forgo something in order to get something and that is the basic rule of sacrifice.

It can be deduced that the work of poverty is to prevent you from living a progressive life and it can only achieve that by releasing strange or difficult forces. Poverty will sacrifice all the time, energy and power just to make sure you do not progress. If poverty can sacrifice to make you poor, why can't you sacrifice to make it rich? It takes sacrifice to fight poverty. Thus, all the forces released by poverty can only be battled with the help of sacrifices. There are number of good sacrifices that one can undertake to fight poverty and these includes;

1. Love
2. Time
3. Money

Love is all about kind feeling, affection, friendship, desire, courtship, fondness, admiration or tenderness. It is a feeling of strong attachment or a serious devotion to another. Love is good but some of the things we love lead to poverty. We desire things which does not add value to our personal life and if we cannot sacrifice these wrong desires, we cannot prosper. You should be able to sacrifice the love you have for that boy or girl who is just wasting your precious time. Do not pay much attention to relationships of no value. Thus, relationships which are likely to destroy your future but you should rather concentrate on positive friendship.

Time is a key factor to prosperity when it is managed well. There are 24 hours in a day, 168 hours in a week and 672 hours in a month. What will happen if all these hours are targeted in achieving success? Success will have no excuse than to come out and the pressure alone will force it out. Most people spend more than half of their time on things which does not add value to personal development. So at the end of the day, you will realize that instead of spending 24 hours to achieve a purpose, you have spent even more than that achieving nothing. It is advisable to sacrifice most of your time on things which will lead you to success and not things which will lead you to poverty.

Money is wealth and wealth is money and it is one of the difficult thing to sacrifice. You can be rich by giving out riches and you can be poor by giving out poverty, for whatever you give so shall you receive. If you want to see money, seed money and if you want to gain money, give money. This basic sacrifice rule will surely bring success.

THE MYSTERY

CHAPTER 8

Mystery is a profound secret which cannot be explained neither can it be understood. It is always beyond human comprehension but in this chapter, I will try to explain and you will have to try to understand. The mysterious nature of poverty beats human understanding and it takes insight to sight such mysteries.

Everything about the world is fifty percent physical and fifty percent spiritual. The physical is what we see, feel and experience and the spiritual is what manipulates the physical. Thus, the spiritual world controls the physical world and anything that has to do with the physical is subordinated to the spiritual realm.

Mysteries do not normally happens in the physical but it is rather connected to the spiritual. Its result is what we see in the physical. There are number of mysteries connected to the spiritual realm which hinder prosperity and produce poverty. These include;

1. **Sexual Intercourse** – sexual intercourse is a beautiful thing designed for married couples in order to create other humans. It is not designed for children and unmarried people. A lot of people have taken sexual intercourse as a fashion. They rock it in different styles and end up in poverty. There is something mysterious about sexual intercourse. If sexual intercourse has the power to create human beings, then it also has the power to destroy human beings. Every human has a burden and the moment you engage in sexual intercourse with other humans is the moment you carry their burden. The more you have sexual intercourse with different partners, the more you carry different burdens and the more you carry different burdens, the more you are initiated into poverty. If that burden does not destroy your money, it will destroy your peace. If it does not destroy your peace, it will destroy your career. If it does not destroy your career, it will destroy your marriage. If it does not destroy your marriage, it will destroy your destiny. Abstaining from sexual intercourse will not only protect you from spiritual harm but it will also save you from absolute poverty.
2. **Drug Abuse** – drug abuse is a serious public health problem that affects almost every nation and community in some way. It is the habitual taking of illegal drugs and many people do not understand why or how other people become addicted to drugs. Addiction is a disorder that affects your brain and behavior. When you are addicted to cocaine, heroin, alcohol and other narcotic drugs, you cannot resist the urge to use them. Drug is not bad but its influence on man makes it bad for that matter the drug is bad. There is something called spirit connected to every drug and this spirit is what controls the brain and behavior of man. Most people who take in these drugs lack the ability to control their destiny towards success and no matter how successful you are, you are likely to fail when you are addicted to drugs. If you want to have

total control of your destiny, you have to try to desist from such drugs and your life will be ordered.

3. **Family Background** – every one belongs to a family and he or she has a peculiar family traits. Some people come from family of idol worship, Christianity, Islamic and other religions. Every family has its own belief system and these systems affect individuals positively or negatively. There is a mystery behind every family and whoever is ignorant of these mysteries is ignorant of life. When you study families very carefully, you will realize that each and every family possess a certain kind of trait. Check and study your family, you will see something which is common or runs through all members. Some might be bareness, sickness, marriage problems, money issues, etc. These problems are in the family because of their root and what they are connected to. In some families, when the mother suffers from sickness, the child will also suffers the same and it passes on to grandchild and so on. Some also lack marriage and it is passed on to generation yet unborn. This kind of hereditary do not respect your beauty, handsomeness, stature and intelligence. It will just catch up with you mysteriously and render you ignorant of the whole fact. You can break this kind of altar by standing out. You have to find out the very root of the whole problem which causes poverty. If you cannot find the root of your family's problem, you cannot uproot the root and if you cannot uproot the root, then you have to stand out. Family altars can deal with you drastically if you do not fight against it. Your destiny is in your hands but if you are not careful, the mistakes of your forefathers will change the root of prosperity. That is why it is necessary to identify such mistakes and deal with it drastically before it deals with you. If your family's problem is sickness, you do not wait to fall sick before you go for check-ups. If it is marriage problems, you do not play with ladies heart. If it is money issues, you

do not spend unnecessarily. If it is bareness, you do not cause abortions.

4. **Other Connections** – There are a lot of connections which are capable of bringing out the best in you and others bring nothing but poverty. Many people think they will prosper when they join secret societies but the truth is, most of these societies will only steal your glory and inject poverty. These societies use money as a deceptive tool to lure people into thinking that they have something better to offer. The desires of people have led many to wrong connections. Humans desire to be famous, popular, powerful and rich. All these desires are good but they come with instructions and principles and these instructions or principles will define the ultimate outcome of your success. A lot of societies came to steal, kill and destroy. They are connected to demonic kingdoms and such connection only leads to death. The power to make riches belongs to God and God has deposited such powers in you and if you cannot create wealth with the power of God, then creating riches with different powers is death.

THE SIN
CHAPTER 9

Sin is a moral deficiency in the character, behavior and the personality of an individual which offends God, nature and man or the violation of certain principles of the earth. Sin can be categorize into many as far as legalistic thinking is concerned. It might be sins of actions, neglect, intent, attitude, etc. Thus, stealing, blaspheming, anger, malicious hatred, bitter envy, adultery and failure to do things we know God, nature and man require of us. The kind of sin you commit determines the type of poverty you possess.

All the poverty man suffers from was as a result of a particular sin and all sins committed will surely beget poverty for sin is poor and poverty is a sin. There are a lot of repercussions connected to every sin known and unknown and that makes ignorance the greatest sin of all sins. It is your responsibility to study, research or find out why things are not going the way it should so that you can find the necessary remedies to it. When you stay idle, stand aloof and watch things go astray then it means you have to be accountable for your actions.

Poverty is broad and the most important aspect people take into consideration is money. Money is powerful and not just powerful but extremely powerful. Its purchasing power is beyond imagination and even has the power to break protocol. It is the breath of the earth and the symbol of the rich. Money is the oxygen that sustains man on earth and without money you are like an alien on different planet.

The difference between greatness inside and outside is money for the money is the physical manifestation of power and strength as far as the earth is concerned. There are a lot of great people out there but due to lack of money, their greatness is hidden and covered. You can soar other areas of your life with your greatness but you can never soar the earth with your greatness when you lack money. Money is great and greatness is money only when its manifestation is powered by influence and strength.

The truth is, you cannot go to a shop and use your ideas to buy food items neither can you explore other parts of the world without a cost. Almost everything on earth is about money. The food we eat, the cloth we wear, the education and even the shelter we enjoy are all connected to money. Money is life on earth but not a life after death.

It is therefore a basic rule to have money on earth and if you lack the ability to have, then it means that, you have violated the fundamental principle of the earth which is a sin against the earth and the earth will surely punish you for that. The punishment is clear, if you do not have money, you do not have food, good education, shelter and possibly good health. That is why it is necessary to adhere to such principles of the earth. There are two ways of obeying the rules of money;

1. Positive obedience to money
2. Negative obedience to money

Positive obedience to money is having money through legal and holistic means. It is require of every man to have money and not just any money but legal and holistic money. We know money is a legal tender, used for purchasing goods and services. It is not illegal tender so everything about money should be legal. As a man, you have to work in order to have and sow in order to reap. You cannot sleep the whole day and have an expectation of controlling money. Whatever you sow or put on the ground will surely give birth to its kind. There are many positive ways a man can earn money and some of these are;

1. Building and marketing your talents through manifestation. Every human is born with a star that comes with unique abilities. It is your responsibility to walk in the direction of your star in order to manifest your uniqueness. There are thousands of people out there who are working but it is not

their job and they will never be fulfilled because they are not on their field. If your work on earth is to job then your earnings on earth will be monetized. Build on whatever talent you have, market it to the world and you will attract money.

2. Earn money by cancelling or reducing things you do not use. There are a lot of engagement which does not bring benefits to humans in any way. Most people have opted for subscriptions which does not add value to their personal life and such things only increases expenses. Cancel things that do not add value to your life and never be bothered to chase unnecessary things out of your life. You can also sell unwanted things to generate income. Your expenses determines your money senses and how you spend the little you have will determine how successful you will become.
3. Pick up a part-time job. You can do other things apart from your main daily activities to earn extra income. There are several things one can do to get cash and this includes running errands for people, make something and sell it online or the closest location, online surveys or marketing, recycling for cash, etc. All these activities help you to boost your income level.
4. Save extra cash. A man without the habit of savings can never save the future generation. Most people find it difficult to save because they earn very little but the truth is, no matter how little your income is, you still have something to save. You should always save some cash before spending. In fact, savings demand a certain level of discipline especially when your income level is very low but it is one of the securities that can save you from unforeseen circumstances. You save not because you have but because you want to save and to be saved.

Negative obedience to money is having money through illegal and other criminal means. It is good to have money because money is good but it is not good to have money which does not sound good. There are so many ways of getting money but not all ways seem right.

If it is a sin to be poor does not mean you should sin to be rich. Whatever you do on earth to get money will be justified at the end. Most people commit a lot of crimes because of money. Money should not be the reason why you should sin but it should be the season for your living. A lot of people are living to please others and some are also living to compete with others. People normally think of what others are thinking about them and others also think of what people are busy doing, these are the very reasons why we commit crimes to compete. You will always be in trouble if you want to live your life to please others and if what people think of you is your business then you really have a business. When you negatively obey money because of these factors, you will negatively have troubles. A lot of people are making wrong decisions, studying wrong courses, marrying wrong partners, joining wrong institutions and engaging in all sort of criminal activities just to look good in order to please people. If you cannot live good to please yourself, living good to please others will only amount to self-destruction. It is not a good thing to negatively adhere to money but it is a good thing to positively adhere to money. There are three classes of people on earth as far as money is concerned;

1. People who perceive money to be evil. These category of people will always have an excuse not to have money. They will tell you the reason why you should not have enough money. They have some wild and weird perception about money. They simply see money as a dangerous weapon to fight against and absolutely nothing motivate them to get money.
2. People who perceive money to be good. People in this category also see money to answer all things. They believe in the power of money and even think man cannot do anything without money.
3. People who perceive money to be good and evil. These class of people perceive money to be normal and abnormal. They

know the repercussions of having money through negative means and enjoying money through positive means.

Thus, the way you perceive money is the way money perceives you and no matter the category you belong, you should not allow money to change you negatively but rather increase you positively. There is hope for you to change the world and until you change, the world would not change. The world is waiting for you! Some might be poor financially, physically, spiritually, economically, mentally, etc. but whatever your poverty might be, you should always remember that poverty is a sin, not a guarantee to sin.

ALBERTOLOGY QUOTES

CHAPTER 10

---

1. Material things are good but its goodness should never and ever override the goodness of mankind.
2. Poverty is the state of desperation, inability to possess possessions and the total confusion in the mind of those who lack the ability to change situations.
3. Man is mind.
4. Poverty was decoded after man realized a certain kind of life below ordinary.
5. A confused mind is a poor mind and a poor mind is a confused world.
6. When the mind suffers, the world suffers.
7. The language of a rich man is never understood by a poor man.
8. Every world has a security gate and the only key to unlock such a world is the mind.
9. The mind controls your life and whatever life you live is the result of your mind.
10. A corrupt mind is a poor mind.
11. Poverty does not joke with life but it can joke with your life.
12. The greatest enemy of man is not him but what is planted in him.

13. There is no prosperity without position and there is no poverty without opposition.
14. The type of a man's poverty is determined by the type of mind he possess.
15. Whatever a man builds on earth will surely be inherited.
16. The Earth was created for our future and the legacy of every man is the life he lived for others and for the future.
17. The future generation will live because we lived and for that reason we are mandated to live because they live.
18. We now live in a world whereby it is very difficult for people to think of others and others to think of people.
19. A man in situation is poverty situated.
20. A person who finds reasons to point fingers is the person who has reasons to be poor.
21. A man can waste many years to think of why he is poor and not why he should be rich.
22. Situation in line with wrong attitude is a disaster.
23. The way you respond to situations is the way situations respond to you.
24. Situational poverty has one major purpose and that purpose is to saturate the minds of those who are not well situated.
25. Humans cannot stop certain poverty from happening but can stop certain poverty from ruling.
26. You can cure ailments when there are positive wings to soar.
27. It only takes lack to be able to know likes.
28. We can overcome situational poverty by reducing accommodation and we can reduce accommodation by reducing negativity and we can reduce negativity by increasing positivity.
29. It takes a man with wisdom to manipulate poverty.
30. Wisdom is wealth and a man without wisdom is a man without wealth.
31. Your state of poverty might be someone's state of prosperity.
32. A man in need is poor indeed.

33. Every man needs a man but a poor man needs more than just a man.
34. A man without a competitor is an invincible man.
35. Humans want to run from economic hardship but they are not even prepared to meet an overflow.
36. Poor economy is a poor individual and a poor individual is a poor economy.
37. You can only discover when you have discovered.
38. The use of resources are counted because of discounted discoveries.
39. High productivity is the utmost key to economic freedom and must never be jeopardized.
40. When resources are scarce, it is our choices that can scrap the power of scarcity.
41. Your attitude towards your body is the true reflection of you.
42. Whatever positive or negative thing you consume will automatically have an impact on the body.
43. To eat rich simply means to balance what you eat even if it is small at the right time and at the right place.
44. People normally say water is life yes is true but water is life when it is safe.
45. A person who is emotionally enslaved lacks the ability to taste good and a poor thought has the power to neutralize nutrients.
46. What grows a child is what keeps a man.
47. A primitive culture is a poor culture.
48. I was told culture is love but later discovered it was a culture of word and not a culture of action.
49. Anything which has the tendency to bring out the beast has the power to kill the beast.
50. Whatever a man believes in has the power to behold him.
51. Most humans are only human beings because they have not discovered their spiritual being.
52. Being spiritually rich is not that difficult but it is difficult for the mind that do not perceive them.

53. Our true self can only be found in God which is due to begin and due to end.
54. If you cannot find God, find you and if you cannot find you, find God.
55. When people cannot find God, let them come to you and when they come to you, let them see God because you are the exact replica of God on earth.
56. To meditate is to speak to the inner spirit, to speak to the inner spirit is to discover the inner spirit, to discover the inner spirit is to discover God and to discover God is to represent God."
57. A man who protects his body and ignores his soul is just like a machine without a proper engine.
58. There is no way a man can be spiritually rich without having the Holy Spirit.
59. Politics benefit from everywhere but not everywhere benefit from politics.
60. The world of today has embraced the false nature of politics which is political trickery and this is the very root of political poverty.
61. Politics of trust and hope is a necessity to national development and without these, there is no politics.
62. The promises of political parties these days can build another planet if they are to go according to what they promised.
63. The prime purpose of every society is to come together in order to help each other but the society we see today came to compete, steal and destroy.
64. Societal poverty depicts suicidal individuals within the society.
65. The power of giving is a supernatural rule of nature which propels prosperity and repels poverty.
66. The mind is the center of a man's life.
67. Show me a desperate man and I will show you a poor man.
68. There is power in positivity to overcome negativity to produce prosperity.

69. The strength of every man is in his mind and it is very difficult to live an extraordinary life with an ordinary mindset.
70. In this era of tribulations, we define problems, problems do not define us.
71. A politician contesting for power is different from a politician in government.
72. You can only rate a poor country when you have rated the attitudes and actions of its politicians.
73. Majority of today's politicians are politicians of today not because of today but because of selfishness.
74. A politician of vision might rise to release hope into the future but a politician of mission is already established in the minds of innocent individuals.
75. Poor nations are very familiar with politicians who came to steal and to destroy and not politicians who give hope and a future.
76. The only difference between a vision and a mission politician is that, a mission politician's vision is to miss the future and a vision politician's mission is to mold the future.
77. A religious leader who only represent God or gods and not the people is not a reliable leader.
78. A true religious leader is the one who represent God and the people.
79. The God we serve is a solution and selfless God but most religious leaders who represent God are only solicitors of selfishness.
80. Almost each and every one belongs to a particular religion and that makes it the biggest media of transformation but it is rather unfortunate that the biggest media of transformation has now become the biggest media of confusion and stagnation.
81. Religious leaders misuse the powers of God to the extent that the ordinary individuals see such powers as a way of escaping hard work.

82. It is good to be better than what you are but it is not really good to be better than what you are not.
83. Hard work is one of the keys to prosperity and every nation's success is as a result of its hard working citizens.
84. Instead of seeking strength to enable us work hard, we rather seek powers that can work for us and this is exactly what has led many nations to be poor.
85. Working hard is not about burning your intestines alive, it's all about working right without giving up.
86. Religious leaders are agents of change and that change must be a change of mind, a change of health, a change of laziness, a change to prosperity and not a change to poverty.
87. A teacher is always an instrument of change in the life of the pupils assigned to teach and every action of such a teacher is tantamount to creation or destruction.
88. Poor classroom management is likely to breed poor future leaders and teachers who have to effect desirable changes are causing miserable changes in the life of innocent pupils.
89. Sometimes, pupils who find it difficult to learn or study new things are the very pupils who have the power to reject old things.
90. Some students learn because they just have to learn and not because they have a justifiable reason to change the future.
91. The most fundamental reason for going to school is to discover your hidden talents and how the school is going to help you maximize such talents in order to help your society and the world at large.
92. It is very unfortunate that most poor nations produce paper graduates and not purpose graduates.
93. Students can use three or four years studying why it is good to farm and not why it is good to be a farmer.
94. Students in most developing countries are trained to know and not trained to do and this lack is as a result of negligence on the part of both students and teachers.

95. Parents are the very people who can add or subtract from the future of their children.
96. It is mandated for each and every parent to work hard not because of today's hunger but the hunger for tomorrow.
97. The foundation you create will determine the future building and building which lacks the ability to accommodate riches is as a result of poor foundation.
98. Words are very powerful and not only powerful but powerful to create and to destroy.
99. The type of words you declare over your children will determine the type of world they will build with your words
100. Parents of this age desire luxury lifestyle than healthy lifestyle in the expense of their children's wealthy lifestyle.
101. Laziness is just a mental catastrophe of both physical and spiritual manifestations.
102. Time has a way of changing situations but not when that situation is under the influence of man.
103. If you do not respect time, time will never respect you and if you do not work within time, time will never work with you, this is what breeds poor individuals, organizations and nations.
104. Competition is a good tool for productivity and creativity but not when the contest is opposed with struggle.
105. Selfishness is one of the most dangerous deoxyribonucleic acid (DNA) of humans and it takes only a selfless mind to neutralize such canker.
106. Employees are the key stakeholders in national development and majority of the powers are invested in them to make a change but the only change most of them make is the change to suit themselves.
107. When the task of employments are not taxed, the tag of poverty is rather taxed.
108. Employing someone who does not qualify for the job will only ends up disqualifying your products or services.

109. It is good to have a sensational mind or soul but it is not good to apply such senses on things which does not make sense.
110. Human beings are interrelated and it is absolutely not receptive to be isolated.
111. We need each other to survive on earth, not after death.
112. Most people are agents of poverty in our life because they desire to plant wrong instead of right.
113. Humans are responsible for solving each other's problem and if you cannot solve any problem, it means you are a problem to our problems.
114. When the wrong people are connected to our life, it plough our right and automatically leads us to our plight.
115. Every person should be careful in choosing his or her spouse when it is time for marriage because the kind of person you connect to determines the kind of burden you will carry.
116. We perceive others to be powerful just because we do not want to receive the power to be grateful.
117. Everything we are meant to be has never meant to be because we did not mean it to be.
118. You need focus and determination to achieve an objective and the only way to be focus is to ignore what others are doing.
119. Poverty is a force and any individual who want to fight poverty also needs a force and not just any force but extraordinary force.
120. Your material wealth should reflect your attitudinal wealth and riches without the right attitude is witches.
121. A poor man's desire is to discuss other people's business in the expense of his or her own business.
122. Time can be traded for money but money can never be traded for time.

123. A man who is poor will use all his strength to chase money instead of working within time.
124. A man can only be considered cheap when his strength, time and mind are of small value.
125. No matter the level of your wealth, when your true desire is to compete, then your true character is poverty.
126. Riches do not consist in having more gold and silver, but having gold and silver which do not contradict peace of mind.
127. All the wrong issues connected to your life are bad in nature and all the right issues connected to your life are good in nature for right is rich and wrong is poor.
128. It is easy to do things which are bad because bad things do not add value to life and things which do not add value are the things which does not come with a cost and whatever thing which does not come with a cost is not difficult to do.
129. You do not have to get money, you have to earn it."
130. There is always a price to pay when you want to opt for easy way to get money and that price will only make life and money difficult.
131. If poverty can sacrifice to make you poor, why can't you sacrifice to make it rich?
132. You can be rich by giving out riches and you can be poor by giving out poverty, for whatever you give so shall you receive.
133. If you want to see money, seed money and if you want to gain money, give money.
134. If sexual intercourse has the power to create human beings, then it also has the power to destroy human beings.
135. Every human has a burden and the moment you engage in sexual intercourse with other humans is the moment you carry their burdens.
136. The more you have sexual intercourse with different partners, the more you carry different burdens and the more

you carry different burdens, the more you are initiated into poverty.

137. Drug is not bad but its influence on man makes it bad for that matter the drug is bad.

138. If you cannot find the root of your family's problem, you cannot uproot the root and if you cannot uproot the root, then you have to stand out.

139. The power to make riches belongs to God and God has deposited such powers in you and if you cannot create wealth with the power of God, then creating riches with different powers is death.

140. The kind of sin you commit determines the type of poverty you possess.

141. All the poverty man suffers from was as a result of a particular sin and all sins committed will surely beget poverty for sin is poor and poverty is a sin.

142. Money is the oxygen that sustains man on earth and without money you are like an alien on different planet.

143. The difference between greatness inside and outside is money for the money is the physical manifestation of power and strength as far as the earth is concerned.

144. You can soar other areas of your life with your greatness but you can never soar the earth with your greatness when you lack money.

145. Money is great and greatness is money only when its manifestation is powered by influence and strength.

146. Money is life on earth but not a life after death.

147. There are thousands of people out there who are working but is not their job and they will never be fulfilled because they are not on their field.

148. If your work on earth is to job then your earnings on earth will be monetized.

149. Build on whatever talent you have, market it to the world and you will attract money.

150. Your expenses determines your money senses and how you spend the little you have will determine how successful you will become.
151. A man without the habit of savings can never save the future generation.
152. You save not because you have but because you want to save and to be saved.
153. It is good to have money because money is good but it is a crime to have money which does not sound good.
154. If it is a sin to be poor does not mean you should sin to be rich.
155. Money should not be the reason why you should sin but it should be the season for your living.
156. People normally think of what others are thinking about them and others also think of what people are busy doing, these are the very reasons why we commit crimes to compete.
157. You will always be in trouble if you want to live your life to please others and if what people think of you is your business then you really have a business.
158. If you cannot live good to please yourself, living good to please others will only amount to self-destruction.
159. It is not a good thing to negatively adhere to money but it is a good thing to positively adhere to money.
160. There is hope for you to change the world and until you change, the world would not change.

Made in the USA
Monee, IL
04 November 2026

67381921R00039